P9-DYE-957

Energy from the Sun

by Melvin Berger

illustrated by Giulio Maestro

Everything that grows or moves or changes must have energy. People and plants, stoves and rocket ships, horses and radios — all need "Go-Power." Without energy this would be a still, silent world.

Where does all this energy come from? What makes plants grow, puppies run, planes fly?

In lively fashion, Melvin Berger explains how almost all the energy on earth begins with the sun. A simple experiment dramatizes the importance of solar energy and helps the reader understand the new ways scientists are discovering to capture, store, and use energy from the sun.

Energy from the Sun

THOMAS Y. CROWELL COMPANY

by
Melvin Berger

illustrated by
Giulio Maestro

NEW YORK

LET'S-READ-AND-FIND-OUT SCIENCE BOOKS

AIR, WATER, AND WEATHER

Air Is All Around You
The Clean Brook
Flash, Crash, Rumble, and Roll
Floating and Sinking
Icebergs
North, South, East, and West
Oxygen Keeps You Alive
Rain and Hail
Snow Is Falling
Sunshine Makes the Seasons
Water for Dinosaurs and You
Where the Brook Begins

THE EARTH AND ITS COMPOSITION

The Bottom of the Sea
Caves
Fossils Tell of Long Ago
Glaciers
A Map Is a Picture
Millions and Millions of Crystals
Oil: The Buried Treasure
Salt
Where Does the Garbage Go?
The Wonder of Stones

ASTRONOMY AND SPACE

The Beginning of the Earth
The Big Dipper
Eclipse: Darkness in Daytime
The Moon Seems to Change
Rockets and Satellites
The Sun: Our Nearest Star
What Makes Day and Night
*What the Moon Is Like**

MATTER AND ENERGY

Energy from the Sun
Gravity Is a Mystery
High Sounds, Low Sounds
Hot as an Ice Cube
Light and Darkness
The Listening Walk
Streamlined
Upstairs and Downstairs
Weight and Weightlessness
What Makes a Shadow?

And other books on LIVING THINGS: PLANTS; LIVING THINGS: ANIMALS, BIRDS, FISH, INSECTS, ETC.; and THE HUMAN BODY

** Available in Spanish.*

 Published simultaneously in Canada by Fitzhenry & Whiteside Limited, Toronto. Manufactured in the United States of America.

Library of Congress Cataloging in Publication Data Berger, Melvin. Energy from the sun. SUMMARY: Briefly describes how, either directly or indirectly, the sun is the primary source for most of the energy used by man. 1. Solar energy—Juv. lit. [1. Solar energy] I. Maestro, Giulio. II. Title. TJ810.B47 621.47 75-33310 ISBN 0-690-01056-7 (CQR)

2 3 4 5 6 7 8 9 10

Energy from the Sun

Energy gives you "go-power." Without energy you could not walk or run, or even move. You could not talk or sing, or eat, or play.

Clocks run on energy. Cars, trains, and airplanes move because of energy. Winds blow and rivers flow because of energy. There is energy all around you.

All things that move or grow use energy. You use energy. All plants and animals use energy. They could not live and grow without energy. Machines use energy too.

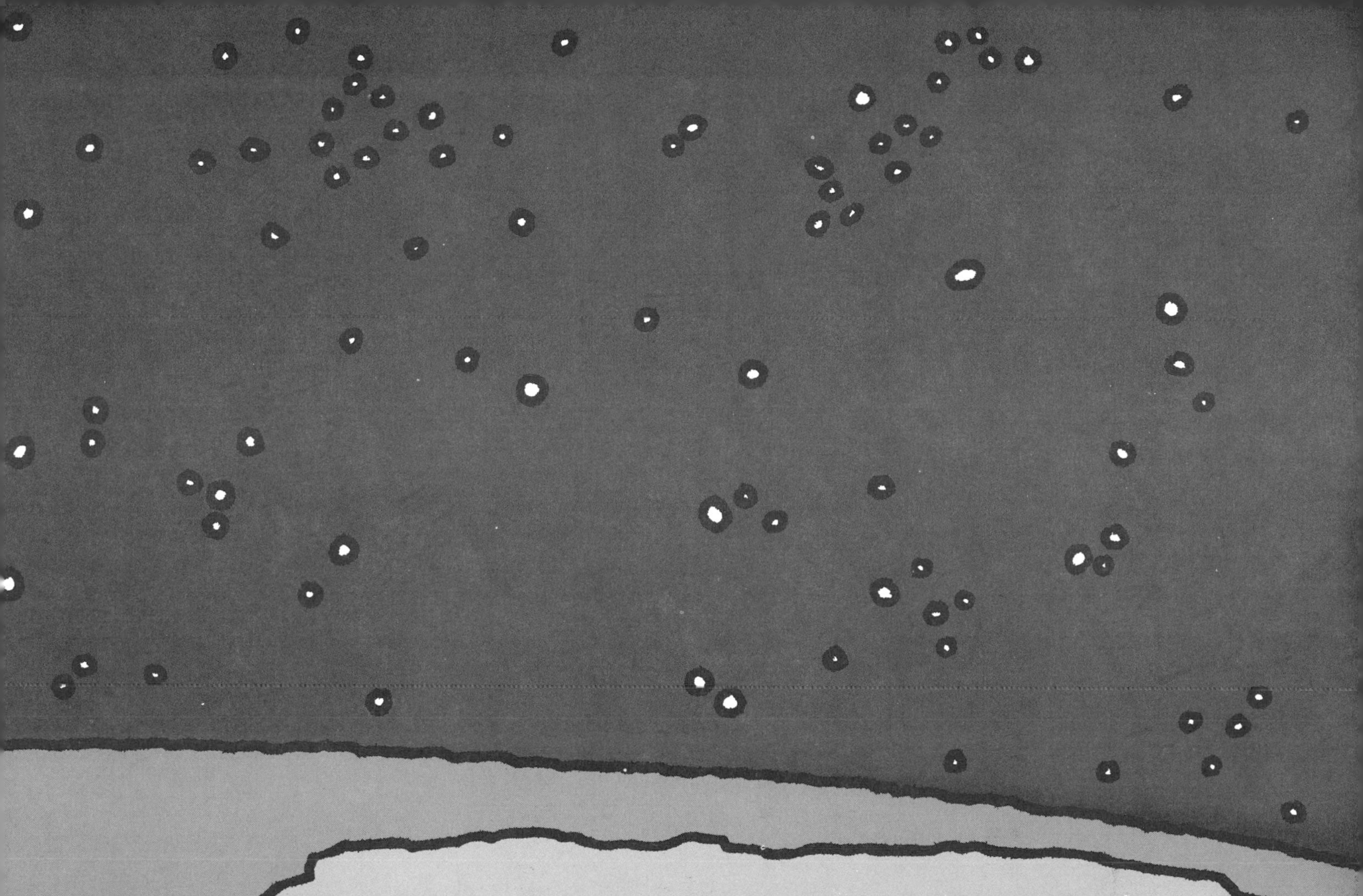

Think what the world would be like without energy.

Nothing would move. Nothing would grow. There would be no light, no warmth, and no sounds. There would be no people, plants, or animals. It would be a dead, still world.

But our world *does* have energy. Most of it comes from the sun. We get heat energy from the sun. We get light energy from the sun.

You can do an experiment to prove that the sun gives heat energy. Fill a jar with cold water. Put it in a sunny place. After a while touch the water. Is the water warm? The heat energy from the sun warmed the water.

Even stones and sidewalks capture energy and store it up. Sometimes sidewalks get so hot you can fry an egg on them.

The energy from the sun helps keep us warm. The energy from the sun also gives us light. But how does the energy from the sun help us move and grow?

The light energy and heat energy from the sun help plants to grow. We eat food that comes from those plants. We also eat food that comes from animals, which eat plants. So we get energy from the sun every time we eat.

Think of some of the foods that you eat.

The peanut butter in your peanut butter and jelly sandwich is made by grinding up the seeds of the peanut plant. The jelly is made from grapes, or strawberries. And the bread is made from flour. Flour is made from wheat, and wheat is the seed of a plant.

A hamburger is made of meat. Meat comes from cattle. Cattle eat grass. And grass is a plant.

Milk comes from cows. And cows, too, eat grass.

Even fish eat tiny water plants.

Think of other foods that you eat. Can you trace all of them back to plants?

Food gives you energy to live and to grow and to move about. But people do not have enough energy to do all of the things they want to do. What if they want to fly to the moon? Or carry a horse? Or cross the ocean?

People use machines to help them do these things. They use rockets to fly to the moon. They use trucks to carry heavy loads. And they use ships to cross the ocean.

These machines need energy. They get the energy from burning fuel. The fuels we use most often are oil, gas, and coal. When fuels burn, they produce heat energy. The heat makes the engines of the machines go.

HORSES

The heat from burning fuel can also be used to make electricity. Electricity is a very useful kind of energy. We cook on electric stoves. We get light from electric bulbs. We listen to radios that run on electricity. Some people even brush their teeth with electric toothbrushes.

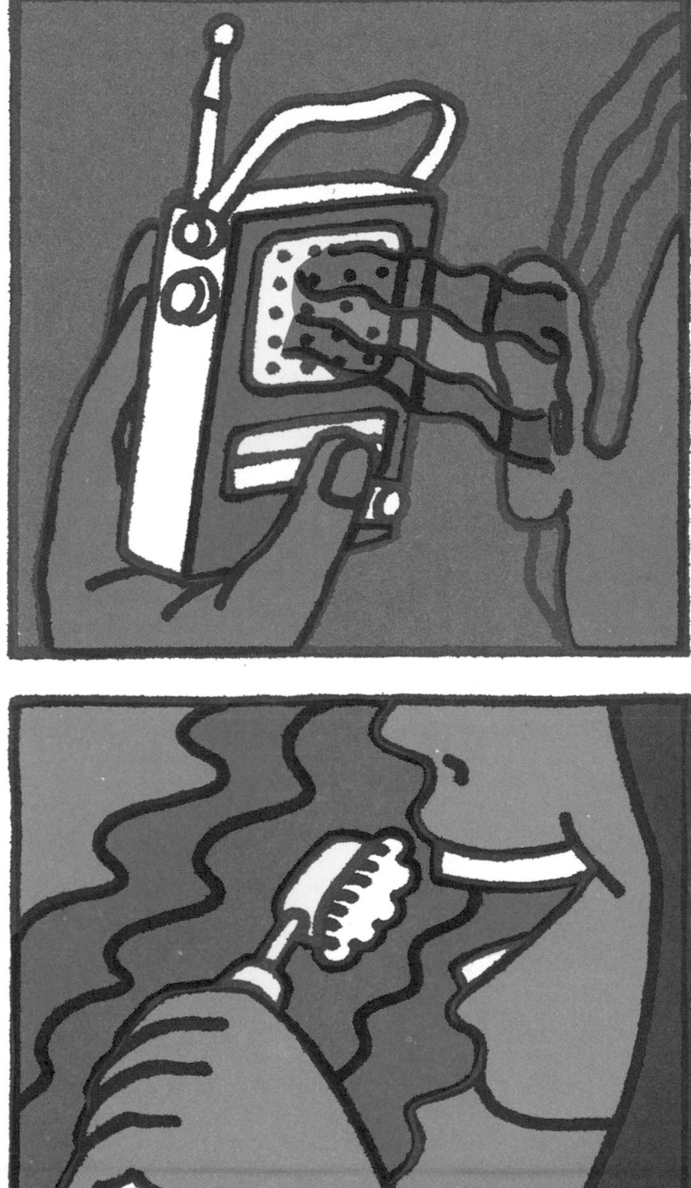

Much of our fuel comes from the earth. We dig wells to get oil or gas. We dig mines to get coal.

How did these fuels get into the earth?

Oil and gas came from tiny plants and animals that lived and died millions of years ago. For thousands of years they were covered with mud, sand, and rock. Over the centuries, these dead plants and animals were formed into thick, gooey layers of ooze. The heat of the earth, and the pressure of the dirt and rock, slowly changed the ooze into oil and gas.

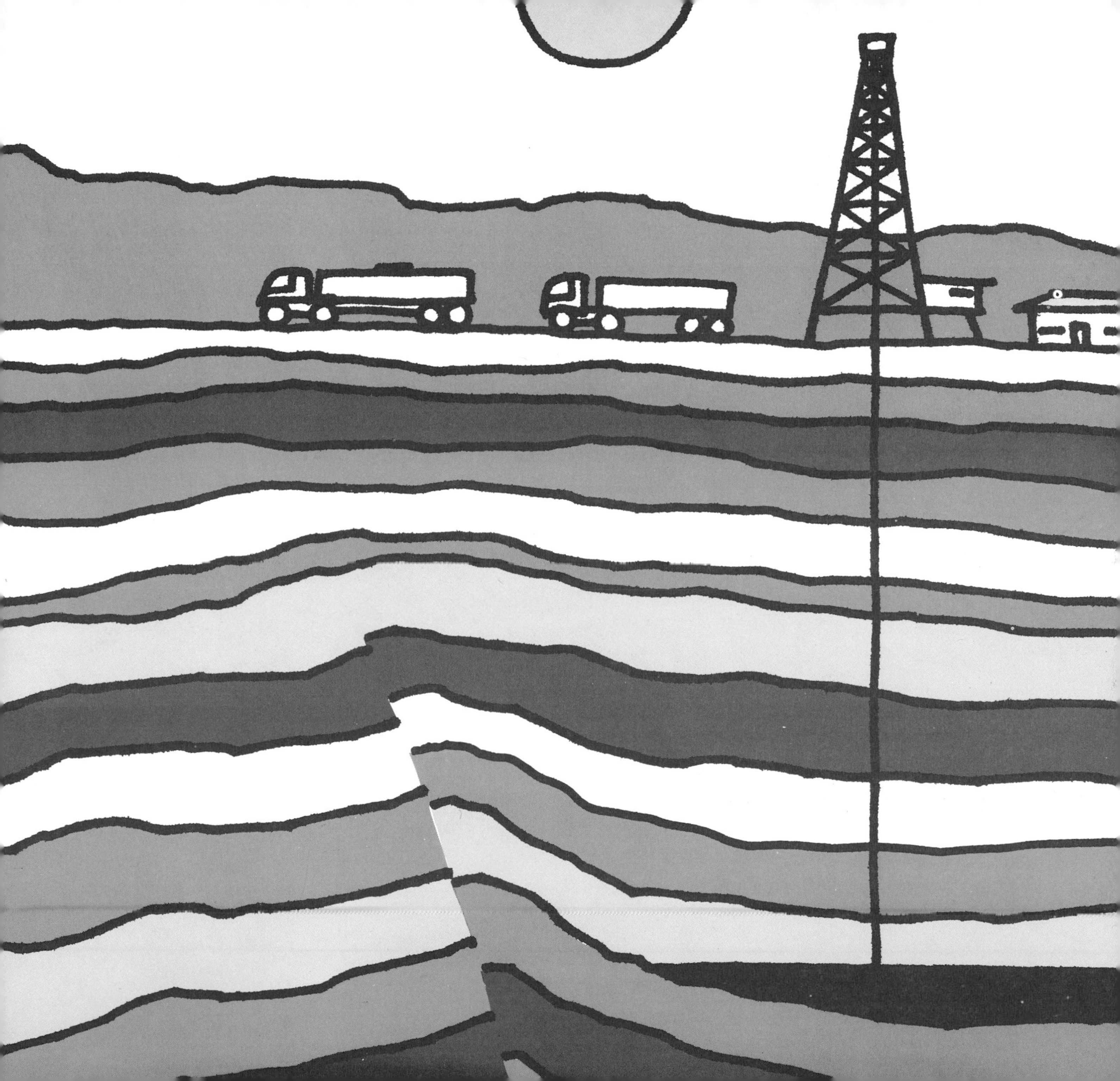

Coal was formed in almost the same way. Coal comes from trees and other plants that lived a long, long time ago. When the trees and plants died they were covered over and pressed into layers of coal.

All our fuels were once living plants and animals. They got their energy to live and grow from the sun—just like the plants and animals that live today.

When we burn fuel today, we are using energy that came from the sun millions and millions of years ago. The sun's energy is still stored in the fuel.

We call the energy that comes from the sun solar energy. Plants trap solar energy. They store it up. When you eat the plants you get the stored-up energy.

Oil, gas, and coal in the earth also have stored-up solar energy. When we burn these fuels we get their solar energy.

But much energy is lost as it passes from sun to plants to fuel to machines. Scientists are now finding ways to use the sun's energy more directly.

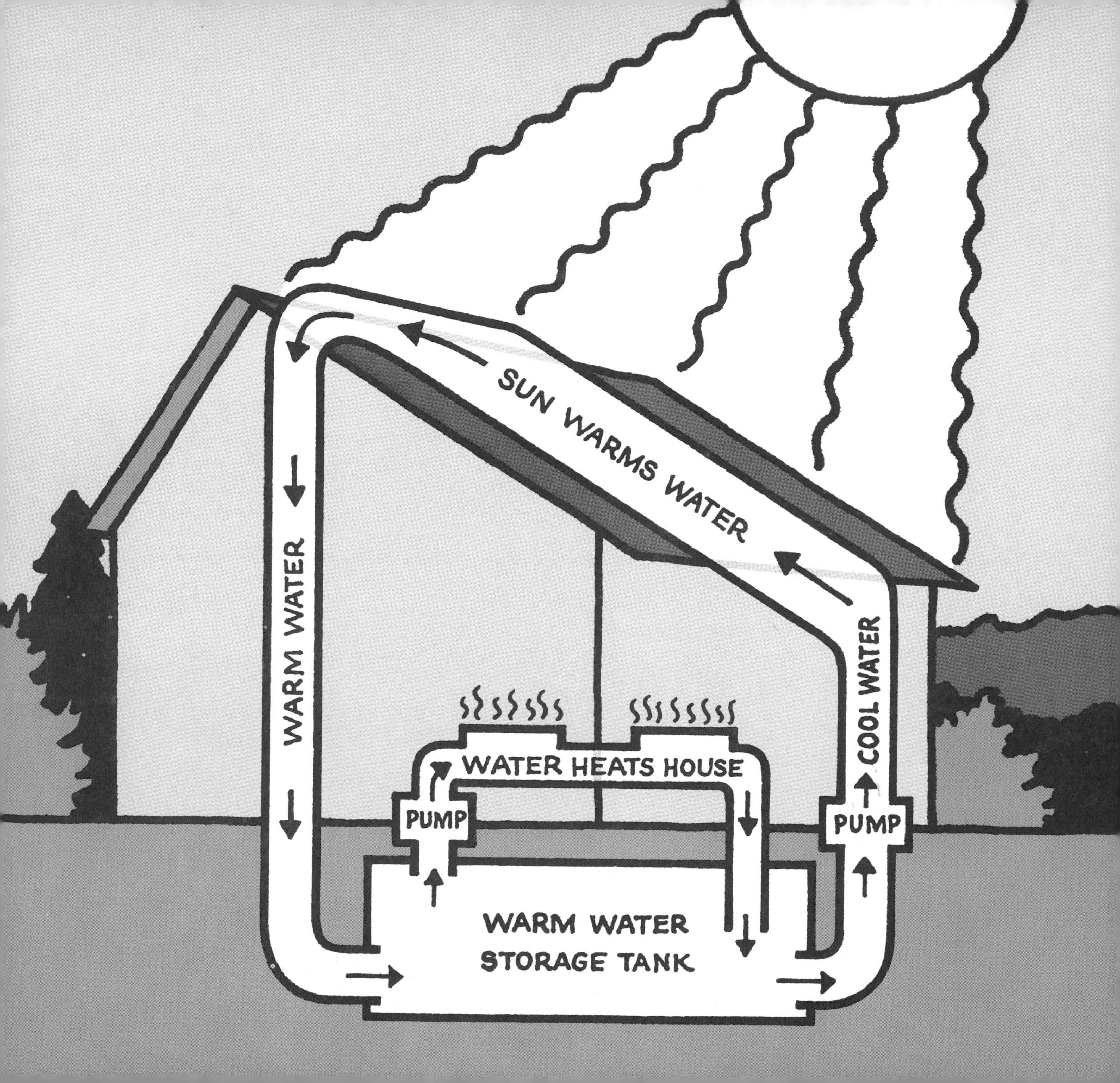
SUN WARMS WATER
WARM WATER
COOL WATER
WATER HEATS HOUSE
PUMP
PUMP
WARM WATER
STORAGE TANK

They are able to heat houses by solar energy. You know from your experiment that solar energy can heat water. The scientists have found how to heat water on the roof of a house. This heat can be stored in the water so the house is warmed day and night.

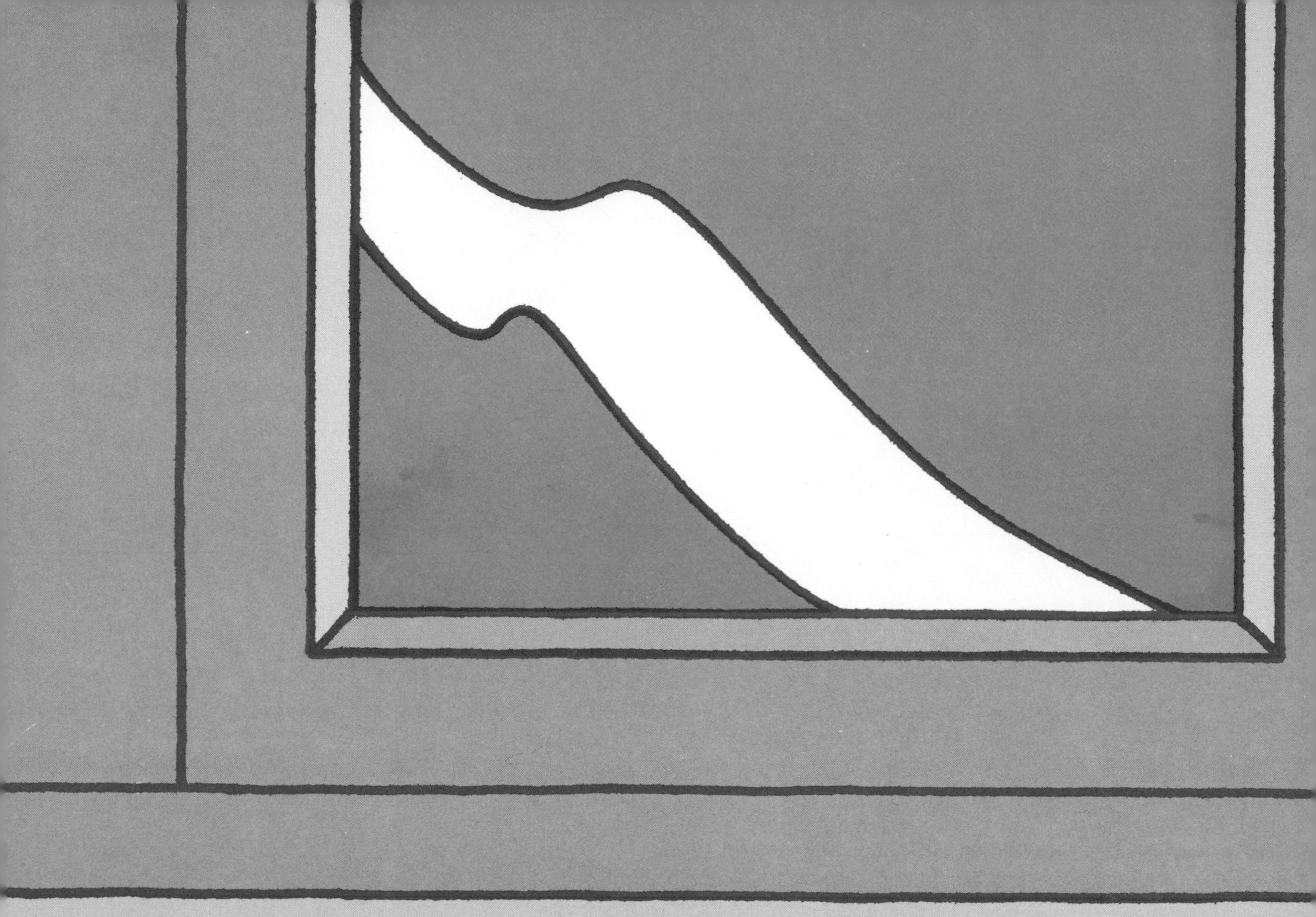

The solar cell is another direct use of solar energy. It uses the light from the sun to produce electricity.

12
9
3
6

Scientists are also looking for other energy sources. Splitting atoms gives us nuclear energy. The blowing wind, the ocean tides, pools of hot water and steam trapped in the earth—these, too, can give us energy.

Sometime in the future we may get most of our energy from one of these sources. But for now and for a long time to come, most of our energy will come from the sun.

Perhaps you will become a scientist. Maybe you will find a new way to capture, store, and use solar energy. Your discovery may help people to live better by using more of the energy from the sun.

About the Author

Melvin Berger feels that it is most important for young people to understand the new world that science is shaping. His many books for young readers are designed to further this understanding. Their subjects range from air and water to computers and enzymes and consumer protection. He is also coeditor of *The Funny Side of Science*, a collection of jokes and cartoons.

Mr. Berger is a native of New York City and holds degrees from the University of Rochester, Columbia University, and London University. He lives in Great Neck, New York, with his wife, Gilda, who is also a writer, and their two daughters.

About the Illustrator

Giulio Maestro was born in New York City and studied at the Cooper Union Art School and the Pratt Graphics Center. He has illustrated many books for children, including several written by his wife, Betsy, a kindergarten teacher. In addition to his picture-book illustration, Mr. Maestro is well known for his beautiful hand lettering and his book jacket designs. He lives in Madison, Connecticut.

Berger, Melvin

Energy from the sun

DATE DUE
